# Freeze Drying or Lyophilization
## An Engineering Perspective

Muhammad Waseem Akbar

Copyright © 2012 Nutritional Guardian

All rights reserved.

ISBN: 9798588946599

# ACKNOWLEDGEMENTS

Freeze drying has many applications in the Food and Pharmaceutical industry. This book is an authentic and supreme emblem of quality that covers the basic concepts related to the lyophilization process. Review of almost all the relevant books and scientific journals is made to make the book error-free. Specifically, this book focusses on the freeze-drying process and their operating parameters for the agricultural products. I want to especially thanks to *Prof. Dr Yongbin Han*, Department of Agricultural Products Processing and Storage Engineering, School of Food Science and Technology, Nanjing Agricultural University, China; to help me to complete this book.

## DEDICATION

This E-Book is dedicated to all Food Engineers and Food Technologist who have worked hard to improve the whole supply chain network of food from farm to fork. Food Engineers and Technologists are working for the previous many years to improve different aspects of food commodities. There is a great role for food professionals in food preservation, food storage, food manufacturing, and food innovation. Food professionals throughout the world have introduced many new innovative techniques to change raw food products into finished ones. So, this book is a dedication to all food professionals around the globe.

## DECLARATION

I hereby declare that this data, neither whole nor as a part has been copied out from any source. It is further declared that I have accompanied the book entirely based on our personnel efforts. If any of this data is proved to be copied out from any source or found to be a reproduction of some other book. We will stand by the consequences. No portion of the work presented has been submitted ever before.

# CONTENTS

|   | Acknowledgements | i |
|---|---|---|
| 1 | Abstract | Pg # 1 |
| 2 | Introduction | Pg # 3 |
| 3 | Chemistry of Agricultural Products | Pg # 10 |
| 4 | Freeze Drying | Pg # 14 |
| 5 | Freeze Drying Process Parameters | Pg # 26 |
| 6 | Thermal Properties of Food | Pg # 33 |
| 7 | Disadvantages | Pg # 39 |
| 8 | Components of a Freeze Dryer | Pg # 41 |
| 9 | Mathematical Modeling | Pg # 44 |
| 10 | Conclusion | Pg # 49 |
| 11 | Nomenclature | Pg # 51 |
| 12 | References | Pg # 56 |

This page is left intentionally blank

# 1 ABSTRACT

This study focusses on the principal steps and parameters involved in the freeze-drying process. Seasonal agricultural products demand preservation and knowledge of spoilage chemistry is important in preservation sciences. Evolution and expansion of lyophilization are very fast as it is unique in drying delicate and Thermo liable products due to low process enthalpy and initial freezing that negligibly shrinks the product structure. Controlling process parameters such as product thickness, chamber pressure and temperature, freezing, primary and secondary drying determines the performance of the freeze-drying. Thermal and freezing product properties include thermal conductivity, thermal diffusivity, specific heat, density and enthalpy that varies the process parameters. Mathematical models are developed to optimize lyophilization process efficiency and reduces associated costs. Inappropriate use of processing parameters causes collapse and disturbs pore paths that degrade product quality. Determination of the end of primary and secondary drying helps to optimize the process and achieves economic sustainability.

**Keywords:** Freeze drying process, modelling, quality, drying kinetics, agricultural products

## 2 INTRODUCTION

Food is consumed for nutritional support that provides energy, growth and therapeutic role to the body. Historically, preservation was practised from the very old times due to the periodical presence of food items and variable production curves [1]. Processing refers to the preservation of food products to get value-added seasonal agricultural products by destroying food spoilage micro-organisms to ensure public health safety [2]. Processed foods are used in food formulations, baby foods, readymade meals, nutraceutical foods, organoleptic foods, special end foods for armies and space missions [3].

Freeze Drying or Lyophilization

Processing of agricultural products can be practised by thermal or non-thermal means. Non-thermal processing uses non-thermic treatments such as high-pressure processing, light (pulsed, ultraviolet), gasses (ozone, cold plasma, chlorine dioxide), chemical (surfactant, chlorine) and ionizing radiation (electron beam, gamma radiation) [4]. Thermal processing uses heat treatment at ambient temperature levels [5] such as baking, ohmic heating, radiofrequency, drying, chilling, frying, microwave, infrared, extrusion, freezing, blanching, pasteurization and sterilization. Food products in Pakistan are mostly thermally processed.

Drying is a coupled heat and mass transfer unit operation in which heat energy diffuses the moisture. Food products are dried to develop new products, to consume seasonal products, and satisfy convenience with retained quality factors that include safety, nutrients, and wholesomeness [6]. Dried fruits are high in nutrients and total solids. Dried products are less in weight and hence easy and cheap to transport [7]. Customer evaluates a product by sensory analysis. Consumer demand processed products that are identical to fresh including durable physiochemical properties [8]. Different types of drying equipment are used such as the sun, foam mat, puff, conveyer, spray, vacuum, drum, radiant heating, continuous infrared, microwave and dielectric heating and tunnel (Oregon, kiln, cabinet, tower, cross flow, centre exhaust, concurrent and counter-current) dryers. Freeze drying can convert these losses to useful products that ensure food security. Freeze drying is the most efficient process in term of its final product [9]. There are different types of freeze-drying equipment such as cabinet or batch, tunnel or semi-continuous, continuous, vacuum spray, coupled dielectric energy and coupled infrared heat energy freeze dryers. Freeze drying is not affected by any climatic conditions such as sun drying. Freeze-drying or lyophilization or cryodesiccation uses the least intense of temperature which

## Freeze Drying or Lyophilization

gives the highest quality end products. It minimally reduces the volume of the product due to initial freezing and sublimation [10]. Freeze drying has very diverse industrial applications and is considered as the best method for dehydrating agricultural (meat, fruits, vegetables, dairy products) and pharmaceutical (protein hydrolysates, blood plasma, vitamins preparation, penicillin, hormones) products [11].

Globally, Pakistan is the sixth most populated country. Pakistan has an agricultural-based economy that provides employment to 42.3% of labour and contributes 18.9% in Gross Domestic Product. 80% of the locals are engaged in agricultural activities [12]. Pakistan is very rich in locally dried fruits market for plums, grapes, almond, fig, walnut, apricot, raisin, dates, pine, coconut, pistachio and cashew nut [13]. Pakistan produces high-quality fresh fruits and vegetables, but export is negligible, due to improper management. The economy of Pakistan can be risen by exporting such value-added products which also creates employment opportunities.

Ancient Egyptian use encapsulation of mummies for preservation. The drying process was used by Samisk and Northern Europeans to preserve their foods [14]. Altmann 1890 [15], firstly developed the systematic applied freeze-drying concept to histological fixation by freezing till −20°C and then applied the subject product to dry. LF Shackell 1909 [16] individually rediscovered freeze-drying after 40 unnoticed years. He uses an electrical pump for creating the vacuum for the biological compounds [17] named as vacuum desiccation of bacteria preservation process and clearly stated the sublimation [18]. Freeze dryer was patent by Tival Henri Louis Paul 1927 [19] and Willaim J. Else 1934 [20] by improving condenser and freezing approach and opened the gateways for industrial applications. Furthermore, Greaves in England, Flosdorf in the United States [21] made a large contribution in freeze-drying. Flosdroff in 1949 [22] published the first book on freeze-drying [23]. Flosdroff in 1952 organized symposia on freeze-drying [24]. The 1950-60s era was the time for the development of freeze-drying [25]. L Rey 1958 in France developed many courses (biennial courses) on freeze-drying [26]. In the 21st century, freeze-drying is used broadly with coupled non-thermal techniques and different lyophilization parameters are improving. Freeze drying is getting modernized and researchers re-investigate the parameters for improving such as rehydration kinetics of strawberries [27], carrots [28] diffusion model of saffron [29], freezing rate effects [30], the water activity of berries and mushrooms [31], process conditions for berries [32], microstructure formation [33], and changes due to freeze-drying [34]. Structural related changes in tissues of apple [35], water activity and glass transition temperature effects optical and mechanical properties [36], structural properties of freeze-dried rice [37], shelf temperature cause pore formation [38], comparison of freeze dryer with other emerging drying techniques [39], coupled infrared radiations and coupled microwave freeze-drying [40], delicate strawberries and mango phenolics freeze-drying [41], and shrinkage porosity due to freeze-drying [42] have also been studied.

# 3 CHEMISTRY OF AGRICULTURAL PRODUCTS

Fruits and vegetables contain 65-95% of moisture content [43]. Perishable agricultural products deteriorate due to their delicate nature. Climacteric fruits continue respiration even after harvesting and spoiled within days due to their increased biochemical changes. Post-harvest losses of fruits and vegetable in Pakistan is 40% [44]. Preservation could defend a huge amount of energy by saving this 40% loss of fruits and millions of value-added products can be manufactured. Food is broadly categorized as plant, ocean and animal source. Food is composed of water,

## Freeze Drying or Lyophilization

carbohydrate, lipid, protein, vitamins, and minerals in variable concentration. Water is a component of all agricultural products and its presence in free form, is the medium for microbial activities [45]. For preservation, spoilage mode of agricultural products is to be controlled. Classification of agricultural products based on their perishability level (stable $<$ 15% MC, semi-perishable $<$ 80% MC, perishable $>$ 80% MC) and pH (high acid $<$ 3.7 pH, acid 3.7 – 4.5 pH, medium acid 4.5 – 5.0 pH, low acid $>$ 5.0) is studied in preservation sciences [46]. The efficiency of food spoilage micro-organisms is maximum, in highly perishable and low – medium acid foods ($>$ 4.5 pH). Stable (legumes, dry cereals, honey, sugar, powdered products, pasta), semi-perishable (potatoes, garlic, ginger, onions, fried snacks), and perishable agricultural products (milk, meat, eggs, fruits, vegetables, cream cakes) have a shelf life of 3-36 months, days-4 months and hours-days respectively. Highly perishable agricultural products need instant drying to get stable [47]. Spoilage agents for agricultural materials are oxidative rancidity, oxidation, enzymatic and the non-enzymatic browning, Millard reaction, caramelization, the non-enzymatic reaction of ascorbic acid and enzymatic rancidity. Furthermore, perishable food products are spoiled and degraded by physical factors, autolysis, micro-organisms, and pests. Mechanical damage caused by birds, rodents, and insects leads to microbial contamination and increases enzymatic activity. Low water activity is required to bound the free available water that provides a medium for spoilage agents. Figure 1 represents the perishability nature that deteriorates the product.

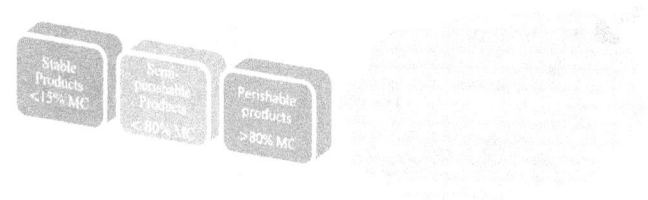

**Figure 1.3:** Perishable nature of the food products leads to

Freeze Drying or Lyophilization

biochemical changes that spoil and creates toxicity

NM Shofian et al. [48] studied the freeze-drying effects on delicate compounds of muskmelon, watermelon, mango, starfruit and papaya. Fresh and freeze-dried samples were examined for total phenolic compounds (TPC), $\beta$-carotene concentration and antioxidant capacity. TPC varies significantly for all tropical fruits excluding muskmelon. Ascorbic acid content remains almost the same. $\beta$-carotene concentration deviates only for watermelon and mango. Anti-oxidant capacity for freeze-dried mango and starfruit was decreased significantly.

# 4 FREEZE DRYING

Drying is carried out to minimize water content in agricultural products. Freeze drying is a preservation technique in which the sample material is firstly frozen below its glass transition temperature and then frozen water sublimes by reducing pressure and providing the low process temperature. Freeze drying is a modern dehydration technique to remove moisture efficiently such that the product suffers less structural deformation. It is a less severe thermal process which minimally degrades the heat-sensitive compounds due to low process enthalpy [49] and protects the Thermo liable constituents. Freeze drying

Freeze Drying or Lyophilization

implements both low and high temperature on the subjected product.

Ayon Tarafdar et al. [50] Quality evaluation was investigated for freeze-dried button mushrooms. Fresh mushrooms having initial $6.14 \pm 0.1$ to $1328.5 \pm 0.1\%$ (d.b) MC with a size diameter of 35-40mm was harvested and stored at $4°C$. The mushrooms were sliced to 0.75cm $\times$ 1.5cm with variable thickness 0.2, 0.5 and 0.8cm and pre-freezes for 24 hours at -22°C. Primary ($-8\ to\ -5°C$) and secondary drying ($25\ to\ -28°C$) both ends when weight becomes constant recorded at a 30min interval and the pressure was maintained at 0.07, 0.04, 0.1mbar for 29 data set values. Optimum values determined were 0.36cm thickness and 0.09mbar pressure. Ascorbic acid, protein, and antioxidant content was measured by standard indophenol, modified Lowry's and DPPH (2,2-diphenyl-1-picrylhydrazyl) assay method respectively. Fresh mushrooms values for antioxidant, protein and ascorbic acid were $9.10 \pm 0.10 mg/g$, $8.43 \pm 0.21 mg/g$, $28.00 \pm 0.53$ $mg/100g$ respectively, compared with freeze-dried samples were $8.60 \pm 0.44 mg/g$, $7.28 \pm 0.56 mg/g$, $26.92 \pm 0.87$ $mg/100g$ respectively. Secondary drying temperature damages the antioxidant and protein content, but ascorbic acid was affected by all process parameters. Effect of temperature was more severe when compared with pressure, which degrades heat-sensitive vitamin C and protein structure. In freeze-drying, contamination can occur from the microorganisms and the equipment material.

Initially, the subjected material is frozen below its eutectic temperature with a refrigeration system. Freezing allows less shrinkage and deformation in structure [51]. Vacuum creation using a vacuum pump by decreasing the pressure below the triple point of water and pressure monitoring during the process with sensors e.g. capacitance manometer. Providing the latent heat of sublimation by

## Freeze Drying or Lyophilization

heating plates. The vacuum pump sucks the sublimated vapours from the process chamber and condensing with a refrigerated condenser to ice. When ice molecules sublime the drying rate slows down because the ice/sublimation front moves backwards. Low pressure or vacuum slows the diffusion rate. The removal of ice leaves a porous structure with negligible volume reduction as compared to other drying methods. Vapours evaporated during drying can be recovered by condensing in a refrigerated condenser [52]. Microscopic porous structure aids reconstitution by regaining maximum water efficiently which enhance product quality, the crispy texture could be recovered, and rehydration cycle time reduces. Freeze-dried products with 2% MC has a shelf life of approximately 2 days. Due to its quality end products the final product doesn't require a refrigeration system. Pore blockage kinetic theory can be studied for improving the quality of dried products if the resistance occurs at the mouth of the pore [53]. The drying process should maximize retention for the delicate components (ascorbic acid, nutrients, vitamins), maintains its nutritional and sensory values and minimizing the processing time and capital cost [54]. Freeze drying is practised for both solid (fruits, vegetables, meat) and liquid raw materials (homogenous solutions). The freeze-drying process is completed in three main stages where raw material pre-treatment is not compulsory. Optimization of the freeze-drying process helps to escalate diffusion rate, reduce process time which lowers energy consumption and hence capital cost. Plackett-Burman experimental design in 1946 [55-58] by RC Plackett and JP Burman and another methodology is Box-Behnken design in 1960 is a useful technique to optimize the freeze-drying process. Steps of lyophilization are listed in the order in figure 2.

Freeze Drying or Lyophilization

**Figure 1.4:** Three steps of lyophilization.

**1.4.1 Raw Material Pre-Treatment:** It is carried out to improve the product quality and to optimize the drying rates by reducing drying cycle time. Pre-treatment includes size reduction, concentration increase, and foaming. A foaming agent is added to food materials that are hard to dry. Djamel Fahloul et al. checks the effects of osmotic dehydration prior to freeze-drying for apricots [59].

**1.4.2 Freezing:** Water in sample foodstuff is converted to the ice by freezing it below its eutectic temperature with a plate, blast, immersion or liquid blast freezing. Successful freezing is considered when 95% of water is converted to ice. Pore size distribution depends on freezing. Selection of optimum freezing rate is compulsory as it disturbs the drying rate and final product quality. Thermal analysis such as cryo-microscopy, differential scanning calorimetry and time versus temperature curve methodologies can be used to identify the eutectic point. The freezing rate determines the growth of crystals and ice morphology. Slow freezing develops bigger crystals, which results in a porous structure, increased drying rate, that may damage the tissues of the product and reduces the rehydration efficiency. Accelerated freezing produces small ice crystals that sponsor intensive nucleation, but easy reconstitution and high drying rates. The optimal freezing rate can be determined by research trials or by reviewing literature for the subjected product. Natalia A. Salazar et al. concluded that higher freezing rate such as $0.4°C/min$ lowers the drying time that reduces process cost up to 30% as compared to $0.1°C/min$ [60].

**1.4.3 Primary Drying**: To sublime ice, the pressure of the process chamber is low (13.5 – 270Pa is absolute) then the vapour pressure of ice in the sample material. Latent heat of sublimation is provided from the heating slabs by conduction and radiation through lower and upper plate respectively to sublime ice. Primary drying or ice sublimation form a dry layer on the top surface of sample product. Diffusion through this partial dry layer or heat movement through frozen and dry layer determines the drying rate. Diffused vapours are condensed by a refrigerated condenser. Non-condensable gasses are removed by a vacuum pump. It is the most time-consuming phase of freeze-drying [61]. The temperature of -20 to +20 is regulated during primary drying. Heat transfer can only be done by conduction or radiation because due to the absence of air convection lacks its medium. The product is heated below its glass transition temperature. Condenser temperature should be lower than the product temperature in primary drying. Maximum heat provided should be below its eutectic temperature that increases the pressure difference between the condenser and the product. Primary drying ends when the product and the shelf temperature become same that stops heat transfer and results in lowering system pressure and condenser temperature values due to no vapours evaporated load. Comparative pressure measurement methods efficiently determine the end of primary drying [62].

**1.4.4 Secondary Drying or Moisture Desorption:** The unfrozen water is removed by diffusion or desorption. For both conduction and radiation through plates during drying forms complex models for the drying rate calculations. But if the heat transfers from one plate only by a single-mode, then the weight change rate can be calculated by the equation (1):

$$\frac{dw}{dt} = \frac{Ak_d(\theta_d - \theta_i)}{L_s l} = \frac{Ab(p_i - p_d)}{l} = A\rho_s(W_0 - W_f)\frac{dl}{dt} \quad (1)$$

## Freeze Drying or Lyophilization

Formulas for both modes of heat transfer are present in the literature [63-66]. Secondary drying begins when the product temperature exceeds its glass transition temperature. Shelf temperature increases by reducing the process chamber pressure to remove the remaining bound water. Maximum allowed temperature is 42°C as protein denatures if the temperature exceeds [67]. End of the drying cycle is 1-3% MC or when the heat and the diffusion transfer stops.

Sakawduan Kaewdam et al. [68] developed a mathematical model for the freeze-drying of mangoes. Quality ripe mangoes with 16-20°Brix were harvested, pre-treated by soaking in 4-5% NaCl solution, peeled, sliced to $3 \times 3 \times 1 cm^3$ and pre-freezes at $-40°C$. Freeze dryer components were the heating plates, condenser, vacuum pump, and the drying chamber. Three primary drying trials were $-40°C$ for 6h, $-20°C$ for 10h, $-10°C$ for 6h and then secondary drying for 2h at 10, 20 and 30°C respectively with 20Pa pressure and 24h total drying time. Water activity, hardness, colour, moisture content, structural characteristics, and effective moisture diffusivity coefficient ($D_{eff}$) were calculated with the water activity meter, texture analyzer, spectrophotometer, infrared moisture determination balance, scanning electron microscope and Fick's diffusion model respectively. Optimum conditions determined were 20Pa, 6h and 10°C for secondary drying. $D_{eff}$ measured was $5.54 \times 10^{-11}$ to $2.90 \times 10^{-10}$ $m^2/s$. Quality parameters measured were hardness 6.1834N, $a_w$ 0.276, and color L*, a*, b* were 79.86, 4.29 and 53.62 respectively. Final MC was 6.8% and SEC was 253.07 $kWh/kg$. Six thin layer models such as Newton (MR = $\exp(-kt)$), page (MR = $\exp((-kt)^n)$), modified page (MR = $a\exp(-kt)$), Henderson-Pabis (MR = $\exp(-kt)$), Logarithmic model (MR = $a \exp(-kt) + c$), and Wang and Sing (MR = $1 + at + bt^2$) were tested for the determination of the coefficient of determination $R^2$, reduced chi-square ($x^2$) and root means square error (RMSE). Modified page model was the most accurate with values of 0.998 $R^2$, 0.00026 $x^2$ and 0.016215 RMSE. The optimum model gives the highest coefficient of determination and the lowest root mean square error values.

# 5 FREEZE DRYING PROCESS PARAMETERS

During freeze-drying, some factors should be critically considered and maintained with high precision. Selection of the operating parameters depends on the subjected product chemistry. Process optimization helps to develop an overall efficient model. Freeze drying is affected by:

**1.5.1 Product Thickness:** Drying time is the function of the sample thickness. By increasing the thickness, more drying time is required hence lowering the drying rate. Thickness recommended by the researchers is 1cm. Drying time can be calculated by equation (2) [69].

$$t = \frac{L^2 \rho (X_o - X_f) \Delta H_s}{8 k_d (T_s - T_{ice})} \quad \text{-------------------- (2)}$$

**1.5.2 Process Temperature and Pressure:** Optimum freezing and drying temperature is necessary to maintain the product quality as the volume reduction is minimum if appropriate temperature and pressure are regulated. Capital cost for fast freezing and drying is low but the product may damage. Operating parameters should avoid the product from chilling injury during freezing and heat burns during drying. The pressure and temperature are the driving forces during drying. The process temperature is controlled by the thermocouples. Jelena Babić et al. [70] studied the effects on chicken (*broiler*) breast meat by varying the operating parameters of freeze-drying. Quality of freeze-dried samples with more thickness was low in term of rehydration as compared to less thickness raw material samples. The author concludes that slow freezing rate had better rehydration %, sensory properties and texture profile. Drying was better at higher pressure (30Pa) than low pressure (25Pa) that also reduces operating cost. Primary drying time depends on the sample thickness, temperature, and pressure.

Primary drying is the longest step of the freeze-drying process, which is responsible for deviating the economy of the process. To make the freeze-drying cost-effective primary drying cycle must be reduced. Increased shelf temperature during secondary drying causes collapses by crossing the critical limits of the collapse temperature and the eutectic melt. Sajal M. Patel et al. [62] worked to investigate the end of primary drying. He concluded that Pirani method is the best to determine the end of primary drying.

If secondary drying initiates before the ending of primary drying, then the sample product exceeds its glass transition temperature and collapse occurs. If secondary drying is delayed, then the freeze-drying process is not optimized and the process cost rises [71].

Chokri Hammami and Frédéric René [72] used a quadratic model or surface response method (RSM) [73] to study the freeze-drying of the strawberry. He concludes that the heating plate temperature and the pressure affect mostly the rehydration ratio, appearance, texture, and the colour as compared to the other parameters.

**1.5.3 Collapse Temperature:** The temperature at which the product structure collapse, volumetric shrinkage occurs and reduced pore size. The plasticizer content and molecular weight affect the collapse temperature and the collapse increases above the glass transition temperature [74]. The process temperature for freezing and primary drying should be below than the collapse temperature, as it results in loss of nutrient, aroma, porosity, structure, colour degradation and increased product density. Collapse temperature values for strawberry, beef, chicken, salmon, orange and guava juice are 70, 60, 60, 60, 49 and 43°C respectively.

**1.5.4 Scorch Temperature:** Scorch temperature should be in its considerable limits otherwise this cause browning during secondary drying. Scorch temperature values for strawberry, beef, chicken, salmon, orange juice, guava juice are $-15, -14, -20, -29, -43$ and $-37°C$ respectively.

**1.5.5 Transition Temperature or Eutectic Temperature:** Temperature at which a product changes from a glassy to a rubbery form or the temperature at which no water molecule remains in the liquid state and converts to ice, so at $T_g$ the product freeze. Glass transition temperature can be determined by equation (3) recommended by M. Gordan-J.S Taylor equation 1952 [75].

$$T_g = \frac{X_1 T_{g1} + kX_2 T_{g2}}{X_1 + kX_2} \quad\quad\quad\quad (3)$$

Where

Freeze Drying or Lyophilization

$$k = C_{p_{glass\ state}} - C_{p_{rubber\ state}}$$

Freeze drying cycle can be optimized by the $T_g$, unfrozen water at $T_g$ and these values can be determined by a freeze-drying microscope or the differential scanning calorimeter by capturing the picture during the collapse, melting, freezing, crystallization and the melt back during drying.

**1.5.6 Chamber Pressure:** Sublimation rate during the primary drying can be calculated by (4):

$$N_w = k_m(P_{ice} - P_{chamber}) \quad \text{---------------------- (4)}$$

The value of $k_m$ depends on how easy the dry layer helps to transfer vapour. Optimum pressure level can be determined by the equation (5) proposed by Tang and Pikal (2004).

$$P_{chamber} = 0.29 \times 10^{0.019T_t} \quad \text{---------------------- (5)}$$

**1.5.7 Heating Plate Temperature:** Adjusting the heating plate temperature is the most time-consuming parameter. During primary drying, the product temperature should be below than the collapse temperature and this can be achieved by lowering the pressure in the process chamber. Secondly, during designing of secondary drying, the dry layer temperature must be lower than the scotch temperature. According to Franks 1990, the product temperature should be greater than $T_g$, to achieve good results. $T_g$ helps to determine the moisture content of the freeze-dried product that helps in the storage level determination.

Specific energy consumption (SEC) depends on the drying time. Water effective diffusion coefficient is directly proportional to the moisture content and becomes constant when the moisture content reduces.

# 6 THERMAL PROPERTIES OF FOOD

Familiarity with thermal properties of subjected material is very important for the effective drying process [76]. Thermal properties include thermal conductivity, specific heat, density, thermal diffusivity and surface heat transfer coefficient. Food drying involves heat transfer. Thermal properties provide the data for the mathematical modelling of the drying process. Properties of the frozen foods include density, thermal conductivity, enthalpy, apparent specific heat and the apparent thermal diffusivity. Thermal properties predict heat transfer rates.

**1.6.1 Thermal Conductivity:** Thermal conductivity is the material ability to conduct heat. Dried products due to their porous structure have low thermal conductivity values. It depends on the process chamber's atmospheric temperature, pressure, surrounding gas nature, porosity, the total solid concentration of the product and the composition of the food commodity [77]. Frozen food can conduct heat four times more as compared to the liquid state product. Thermal conductivity is used in determining the thermal energy transfer values for conduction. It is denoted by $k$. Its unit is $W/mK$. It can be calculated by using equation (6):

$$k = \frac{Qx}{A\Delta T} = \frac{Heat\ Flux}{Temperature\ gradient} \quad \text{-------------------- (6)}$$

Thermal conductivity can be predicted by the Kopleman, Mattea-Urbicain-Rotstein, random parallel, series, Krischer, Maxwell-Eucken, and improved thermal conductivity prediction models [78-81]. It can be measured

## Freeze Drying or Lyophilization

by steady-state (longitudinal heat flow, radial heat flow, the heat of vaporization, heat flux, and differential scanning calorimeter methods) and the unsteady state models (thermal conductivity probe, transient hot-wire, modified fitch, point heat source and comparative methods) [82]. Design of the thermophysical properties is mandatory for predicting the processing time for freezing and drying.

**1.6.2 Specific Heat:** Energy required to raise a unit temperature of a unit mass product. It is denoted by $C_p$. Its unit is $kJ/kg°C$. Siebel 1892 suggests the equation (7) for the measurement of the specific heat for juices and vegetables [83].

$$C_p = 0.837 + 3.349 X_w^w \quad \text{---------------------- (7)}$$

And, for products below their freezing temperature can be determined by equation (8)

$$C_p = 0.837 + 3.349 X_w^w \quad \text{---------------------- (8)}$$

It depends on the mass fraction of water. Specific heat can be predicted by other mathematical methods developed by Heldman 1975, Choi and Okos 1986, Rahman 1992 and Riegel 1992. It can be measured by the mixture, guarded plate, comparison calorimeter, adiabatic agricultural calorimeter, Gorden and Thorne 1990 and the differential scanning calorimeter methods [82].

**1.6.3 Thermal Diffusivity**: Material ability to transfer heat with respect to the store the thermal energy [84]. Thermal diffusivity of the frozen food product is higher as compared to the unfrozen water in the food product [85]. Enthalpy at $-40°C$ is zero and with the increase of temperature the enthalpy increases. It is denoted by $\alpha$. Its unit is $m^2/s$. It can be calculated by equation (9):

$$\frac{\alpha_1}{\alpha_2} = \frac{\Delta t_1}{\Delta t_2} \quad \text{---------------------- (9)}$$

It can be measure by the direct (by using density, thermal

conductivity and specific heat) and the indirect methods (temperature history, thermal conductivity probe and Dickerson methods). Gorden and Thorne 1990 develop a method for the measurement of the thermal diffusivity of foods [86].

**1.6.4 Density:** Ratio of mass to volume. It is denoted by $\rho$. Its unit is $kg/m^3$. The density of gasses varies with varying temperature. For gasses, $\rho$ decreases when the temperature rises and with the pressure increase density of the gasses also increases [87]. It can be measured by the pycnometer, hydrometer and predicted by the Choi and Okos 1986 equations that are temperature-dependent because density is temperature-dependent. A frozen food product containing ice has less density than the food products containing water in the liquid state.

**1.6.5 Enthalpy:** It is the total heat content or energy level of a body and is the combination of the internal energy into the pressure and volume. It is denoted by H. Its unit is $J/kg$. It is represented by equation (10).

$$H = P + V \quad \text{----------------------} \quad (10)$$

**1.6.6 Latent Heat:** Heat gained or released at constant temperature and pressure during the phase change of a unit mass product.

**1.6.7 Surface Heat Transfer Coefficient:** It depends on the surface texture, shape, temperature difference, and fluid properties. During drying both latent and sensible heats are removed from the product.

## 7 DISADVANTAGES

Removal of water makes the high concentration and pH change which may denature the protein. The capital cost of the freeze dryer is high, and the processed material is hygroscopic in nature. Freeze dryer needs the high installation cost, requires skilled labour that also adds cost to the process and overall, the process consumes high content of energy in freezing and the drying. The freeze-drying process requires approximately 9 – 30 hours that increases cost also. Table 1 represents the cost per utility of the freeze-drying process.

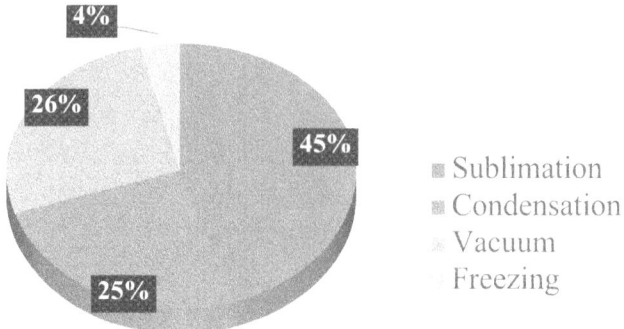

**Table 1:** Energy consumption per utility during lyophilization.

# 8 COMPONENTS OF A FREEZE DRYER

Freeze dryer consists of the vacuum drying chamber, vacuum pump, Ice condenser, shelves for heating, rubber valves and manifolds [88]. Different freeze dryer is present in literature such as freeze dryer with expendable refrigerant, microwave freeze dryer by D Fissore 1971 et al., K Witkiewicz 2010 et. al., H Jiang 2010 et al [88-90] and many more are present in the literature. Figure 5 represents the freeze dryer used by Duan et al. for the study of browning behaviour of mushrooms [91]. Jiapeng Huang and Min Zhang 2016 asses the quality characteristics of okra and the equipment was explained by figure 6 [92].

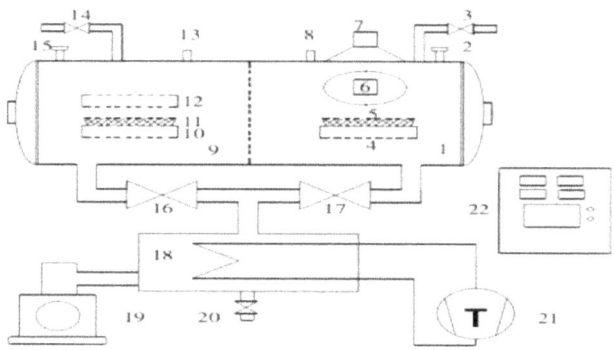

**Figure 6:** The VFD system consists of 1, Microwave freeze-drying chamber; 2, optical fibre temperature sensor; 3, vacuum breakage valve for MFD; 4, the sample supporting plate; 5, MFD sample; 6, 7, microwave source; 8, pressure sensor for MFD chamber; 9, freeze-drying chamber; 10, 12, heating plate; 11, VFD sample; 13, pressure sensor for VFD chamber; 14, vacuum breakage valve for VFD; 15, temperature sensor; 16, VFD vacuum valve; 17, MFD vacuum valve; 18, cold tarp; 19, vacuum pump; 20, drainage valve; 21, refrigeration compressor; 22,

Freeze Drying or Lyophilization

control system. [92].

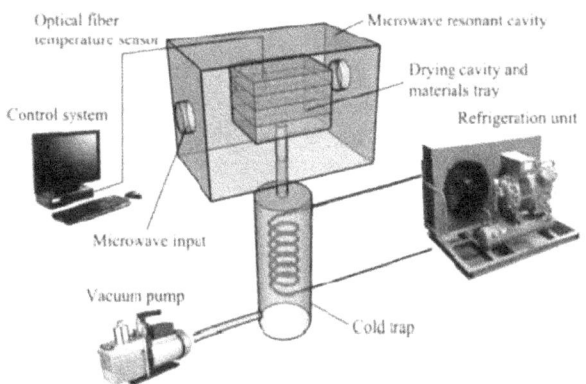

**Figure 5:** A microwave-assisted freeze developed by Duan et al. 2015.

# 9 MATHEMATICAL MODELING

James P.George and A.K.Datta developed the heat and mass transfer model in equation (11) of the vegetable slices for the effective internal mass transfer coefficient ($D'$) [93].

$$D' = \frac{(1-x)^2 RTL^2 X_o \rho_s}{2tM_w(\rho_{fw} - \rho_{ew})} \quad \text{-----(11)}$$

Atmospheric freeze-drying models have been developed for the heat and mass transfer equations [94] by Heldman and Hohner 1974 in equation (12) and (13) respectively

Freeze Drying or Lyophilization

[95].

$$\rho(C_p + MC_{p,w})\frac{\partial T}{\partial \theta} = \frac{\partial}{\partial \eta}\left(\frac{\lambda}{D_{fg}}\frac{\partial T}{\partial \eta}\right) + \frac{D_e}{D_{fg}}C_{p,w}\frac{\partial P'}{\partial \eta}\frac{\partial T}{\partial \eta} + \rho \Delta H_v \frac{\partial M}{\partial \eta}$$

-------------------- (12)

$$\left[\frac{\varepsilon M_w}{RT} + \frac{\rho}{P_{sat}}\frac{dM}{d\varphi}\right]\frac{\partial P'}{\partial \theta} = \frac{\partial}{\partial \eta}\left(\frac{D_e}{D_{fg}}\frac{\partial P'}{\partial \eta}\right)$$

-------------------- (13)

AFD in a fluid bed for an absorbent is given by the Wolff and Gibert 1990 equation labelled as equation (14) [96].

$$\frac{A(P'_{sat}P'_c)}{R_e + R_i x} + \left[\frac{m_a}{\frac{1}{a_w}\frac{da_w}{dY_a} - B_r \log\left(\frac{a_{w,a}}{a_w}\right)}\right] \times \left(\frac{1}{P'_c}\frac{dP'_c}{dt} - \left[\frac{A_r \exp(-B_r Y_a) + \frac{\Delta H_v M_w}{R}}{T^2}\right]\frac{dT}{dt}\right) = 0$$

------------------- (14)

The variables $R_e$ and $R_i$ can be calculated by the equations (15) and (16) respectively.

$$R_e = \frac{1}{\beta_{ext}} + \frac{1}{\alpha_{ext}}\frac{(\Delta H_s)^2 P'_{sat} M_w}{RT^2}$$

-------------------- (15)

$$R_i = \frac{RT}{M_w D} + \frac{1}{\lambda_s} \frac{(\Delta H_s)^2 P'_{sat} M_w}{RT^2} \quad \text{----------------- (16)}$$

Wei-Youh Kuu et al. [97] determined the mass transfer coefficient by estimating the parameters and modelling of the dried layer during the primary drying by using MJ Pikal 1983 heat and mass transfer equations. Wei Wang et al. [98] developed a model for hygroscopic material skim milk. The model was numerically solved. Modelling help in the process and the equipment designing, analyzing diffusion mechanics, to locate the optimal conditions and to calculate the process variables that are difficult to know by experimental methods. Drying models are based on AV Luikov's 1975 system [99] and Whitaker 1977, 1980 [100-102] theory. The raw material for the drying process is categorized as hygroscopic that is filled with water and non-hygroscopic materials which contain negligible bound water. J.F. Nastaj and K. Witkiewicz developed a model for the primary and secondary drying during freeze drying.

## Freeze Drying or Lyophilization

Simulation and experimentally performed results when compared, showed good results [103]. Drying is a thermal process. Values of convective heat transfer coefficient and thermal diffusivity are calculated variables, for the mathematical simulation and determining these factors could optimize the simulation results [104]. Different mathematical modelling was made and found in the literature on primary drying modelling, lyophilization optimization using modelling, couple radiation freeze dryer, scale-up issue during freeze-drying, drying steps modelling for lyophilization, pressure rise analysis model, multi-dimensional lyophilization by A Giordano 2011 et al., R Pisano 2011 et al., R Pisano 2008 et al., S Rambhatla 2003 et al., P Sheehan 1998 et al., P. Chouvenc 2004 et al., K Nakagawa 2015 et al. respectively. Aroldo Arévalo-Pinedo and Fernanda E.X. Murr researched on the drying kinetics of the pumpkin. He determined that freezing was better than blanching for a primary drying process. Best drying curves were obtained by Fick's $2^{nd}$ law of diffusion

## Freeze Drying or Lyophilization

1855 and diffusion rate can be enhanced by increasing the drying temperature [105]. Different models can predict freezing time such as heat transfer models (numerical, analytical and dimensionless variables methods) and heat and mass transfer models. Prediction of the transfer coefficient during drying depends on the heat and mass transfer coefficients and thermophysical properties [106]. Due to the high processing cost associated with the freeze-drying requires careful process control and attention [107]. Different researches worked on the freeze-drying modelling such as Antonio Vega-Gálvez et al. and made a model for the mass transfer process during reconstitution of Aloe vera (*Aloe barbadensis Miller*) by using Peleg and Weibell, and Fick's model [108] and the investigation of the freeze-drying processes such as microwave-assisted freeze-drying [109].

# 10 CONCLUSIONS

Freeze-drying retains the quality of the agricultural products in term of nutritional value, heat-sensitive compounds, colour, shape, texture, size, aroma, and flavour. Ongoing development of freeze-drying and future aspects makes its application very feasible. Principal process parameters vary with subjected products chemistry. Freeze drying is an old and evolutionary technique throughout the years and improving every day. But freeze drying is associated with the high process and fixed cost. Freeze dryer firstly freezes the water in the product and then providing the latent heat of sublimation

## Freeze Drying or Lyophilization

to sublimate frozen water and then increases the heat to remove remaining bound water.

# 11 NOMENCLATURE

| | |
|---|---|
| MC | Moisture content |
| MFD | Microwave freeze-drying |
| VFD | Vacuum freeze-drying |
| $A$ | Exchange surface or drying area $(m^2)$ |
| $A_r, B_r$ | Constant |
| $a_w$ | Water activity |
| $a_{w,a}$ | The water activity of the absorbent |

# Freeze Drying or Lyophilization

| | |
|---|---|
| $b$ | Dry layer permeability to water vapours |
| $C_p$ | Specific heat of dry product at constant pressure (J/kgK) |
| $C_{p,i}$ | Specific heat of ice at constant pressure (J/kgK) |
| $C_{p,w}$ | Specific heat of ice at constant pressure (J/kgK) |
| $D$ | Dry layer water vapour diffusivity $(m^2/s)$ |
| $D_e$ | Effective water vapour diffusivity $(m^2/s)$ |
| $D_{fg}$ | Water vapour diffusivity for free gas (air + vapor) $(m^2/s)$ |
| $\dfrac{dl}{dt}$ | Rate of change of thickness for the dried layer |
| $\dfrac{dw}{dt}$ | Rate of weight change |
| $H$ | Enthalpy (J) |
| $\Delta H_s$ | Latent heat of sublimation (J/kgK) |
| $\Delta H_v$ | Latent heat of vaporization (J/kgK) |
| $k_d$ | Thermal conductivity of the dried layer (W/mK) |
| $k$ | Thermal conductivity (W/mK) |
| $k_m$ | Coefficient |
| $L_s$ | The heat of sublimation $(K)$ |
| $l$ | Dry layer thickness $(m)$ |
| $L$ | Slab thickness $(m)$ |
| $L'$ | Sample half-thickness $(m)$ |

Freeze Drying or Lyophilization

| | |
|---|---|
| $M$ | Dry basis moisture content |
| $M_w$ | The molar mass of water $(kg/mol)$ |
| $m_a$ | Mass of water in the absorbent $(kg)$ |
| $N_w$ | Sublimation rate |
| $P_{chamber}$ | Process chamber pressure $(Pa)$ |
| $p_i$ | Ice front water vapour pressure $(Pa)$ |
| $p_d$ | Dry layer water vapour pressure $(Pa)$ |
| $\rho_s$ | Dry layer density $(kg/m^3)$ |
| $P'$ | The partial pressure of water vapour $(Pa)$ |
| $P'_c$ | The partial pressure of water vapour in the drying chamber $(Pa)$ |
| $P'_{sat}$ | Water vapour saturation pressure $(Pa)$ |
| $P$ | Pressure $(Pa)$ |
| $\rho$ | Solids bulk density $(kg/m^3)$ |
| $Q$ | Heat flow $(W)$ |
| $R$ | Molar gas constant $(J/Kmol)$ |
| $R_e$ | External combined heat and mass transfer resistance $(Ns/kg)$ |
| $R_i$ | External combined heat and mass transfer resistance $(Ns/kg)$ |
| $T_g$ | Glass transition temperature $(K)$ |

Freeze Drying or Lyophilization

| | |
|---|---|
| $T_{g1}$ | The glass transition temperature of dry solids ($K$) |
| $T_s$ | Maximum surface temperature ($K$) |
| $T_{ice}$ | Maximum ice temperature ($K$) |
| $T_t$ | Target temperature ($K$) |
| $T$ | Temperature ($K$) |
| $t$ | Time ($s$) |
| $\dot{t}$ | Drying time ($s$) |
| $\Delta T$ | Temperature difference ($K$) |
| $V$ | Volume ($m^3$) |
| $W_0$ | Initial MC of the subjected material |
| $W_f$ | Final MC of the subjected material |
| $X_1$ | Mass fraction of dry solids ($kg/kg$) |
| $X_2$ | Mass fraction of water ($kg/kg$) |
| $X_o$ | Initial MC ($kg/kg$) |
| $X_f$ | Final MC ($kg/kg$) |
| $x$ | Distance ($m$) |
| $x_1$ | Product thickness ($m$) |
| $X_w^w$ | Mass fraction of water ($kg/kg$) |
| $Y_a$ | The absolute humidity of the absorbent |
| $\theta_d$ | Dried layer top surface temperature ($K$) |
| $\theta_i$ | Ice front temperature ($K$) |

Freeze Drying or Lyophilization

## Greek Symbols

| | | |
|---|---|---|
| $\alpha_{ext}$ | External heat transfer coefficient | $(W/m^2K)$ |
| $\beta_{ext}$ | External mass transfer coefficient | $(W/Ns)$ |
| $\varepsilon$ | Porosity | |
| $\eta$ | Dimensionless distance | $(s/L)$ |
| $\theta$ | Dimensionless time | $(D_{fg}t/L^2)$ |
| $\lambda_s$ | Dry layer thermal conductivity | $(W/mK)$ |
| $\rho$ | Bulk density of the porous zone | $(kg/m^3)$ |
| $\varphi$ | Relative humidity | |

# 12 REFERENCES

1. Fellows, P.J., *Food processing technology: principles and practice*. 2009: Elsevier.

2. Aguilera, J.M. and D.W. Stanley, *Microstructural principles of food processing and engineering.* 1999: Springer Science & Business Media.

3. Ratti, C., *Hot air and freeze-drying of high-value foods: a review.* Journal of food engineering, 2001. **49**(4): p. 311-319.

4. Worobo, L.F.P.a.R., *Non-thermal or Alternative Food Processing Methods to enhance Microbial Safety and Quality.* 2011.

5. Food, S., *Thermal Processing of Food.* 2014. p. 4-21.

6. Moazzam Rafiq Khan, M.A.R.a.M.A.S., *Processing of fruits and vegetables*, in *Handbook of food science and technology*, I.A.K.M. Farooq, Editor. 2017, University of Agriculture Faisalabad: University of Agriculture Faisalabad. p. 145-164.

7. Sagar, V. and P.S. Kumar, *Recent advances in drying and dehydration of fruits and vegetables: a review.* Journal of food science and technology, 2010. **47**(1): p. 15-26.

8. Rangel-Marrón, M., et al., *Estimation of moisture sorption isotherms of mango pulp freeze-dried.* International Journal of Biology and Biomedical Engineering, 2011. **5**(1): p. 18-23.

9. Awan, J.A., *Food Processing and Preservation.* 2011, Faisalabad: Unitech Communications.

10. Liapis, A.I. and R. Bruttini, *Freeze-drying*, in *Handbook of industrial drying.* 2006, CRC press. p. 282-309.

11. Ciurzyńska, A. and A. Lenart, *Freeze-drying-application in food processing and biotechnology-a review.* Polish Journal of Food and Nutrition Sciences, 2011. **61**(3): p. 165-171.

12. Pakistan, G.o., *Economic Survey of Pakistan.* , F. Department, Editor. 2017-2018, Government of Pakistan: Islamabad. p. 13-32.

13. Aazim, M., *Production and exports of fresh and dry fruits*, in *Dawn.* 2016, Dawn News: Karachi.

14. Corver, J., *The Evolution of Freeze-Drying.* Innovations in Pharmaceutical Technology, 2009: p. 66-70.

15. Gersh, I., *The Altmann technique for fixation by drying while freezing.* The Anatomical Record, 1932. **53**(3): p. 309-337.

16. Shackel, L., *A improved method of desiccation with some applications to biological problems.* Amer J. Physiol, 1909. **24**: p. 325-340.

17. Shackell, L., *An improved method of desiccation, with some applications to biological problems.* American Journal of Physiology-Legacy Content, 1909. **24**(3): p. 325-340.S

18. Hammer, B.W., *A note on the vacuum desiccation of bacteria.* The Journal of medical research, 1911. **24**(1): p. 527.

19. Paul, T.H.L., *Means for preparing products of organic origin.* 1927, Google Patents.

20. Elser, W.J., *Method of desiccating liquids and semisolids.* 1934, Google Patents.

21. Meryman, H., *Historical recollections of freeze-drying.* Developments in biological standardization, 1976. **36**: p. 29-32.

22. Rey, L. and J.C. May, *Freeze-Drying/Lyophilization Of Pharmaceutical & Biological Products, Revised and Expanded.* 2004: CRC Press.

23. Couriel, B., *Freeze-drying: past, present, and future.* Journal of the Parenteral Drug Association, 1980. **34**(5): p. 352-357.

24. Flosdorf, E.W., *Freeze-drying. Drying by sublimation.* Freeze-drying. Drying by sublimation., 1949.

25. Gronka, P., *Freeze-dry history and process.* 2014.

26. Rey, L.R., *Thermal analysis of eutectics in freezing solutions.* Annals of the New York Academy of Sciences, 1960. **85**(2): p. 510-534.

27. Vergeldt, F., et al., *Rehydration kinetics of freeze-dried carrots.* Innovative Food Science & Emerging Technologies, 2014. **24**: p. 40-47.

28. Meda, L. and C. Ratti, *Rehydration of freeze-dried strawberries at varying temperatures.* Journal of Food Process Engineering, 2005. **28**(3): p. 233-246.

29. Acar, B., H. Sadikoglu, and I. Doymaz, *Freeze-Drying Kinetics and Diffusion Modeling of Saffron (Crocus sativus L.).* Journal of Food Processing and Preservation, 2015. **39**(2): p. 142-149.

30. Ceballos, A.M., G.I. Giraldo, and C.E. Orrego, *Effect of freezing rate on quality parameters of freeze-dried soursop fruit pulp.* Journal of food engineering, 2012. **111**(2): p. 360-365.

31. Khalloufi, S., J. Giasson, and C. Ratti, *Water activity of freeze-dried mushrooms and berries.* Canadian Agricultural Engineering, 2000. **42**(1): p. 51-56.

32. Reyes, A., et al., *Comparative study of different process conditions of freeze-drying of 'Murtilla' berry.* Drying Technology, 2010. **28**(12): p. 1416-1425.

33. Harnkarnsujarit, N., S. Charoenrein, and Y.H. Roos, *Microstructure formation of maltodextrin and sugar matrices in freeze-dried systems.* Carbohydrate polymers, 2012. **88**(2): p. 734-742.

34. Oikonomopoulou, V.P., M.K. Krokida, and V.T. Karathanos, *The influence of freeze-drying conditions on microstructural changes of food products.* Procedia food science, 2011. **1**: p. 647-654.

35. Venir, E., et al., *Structure related changes during moistening of freeze-dried apple tissue.* Journal of food engineering, 2007. **81**(1): p. 27-32.

36. Moraga, G., et al., *Implication of water activity and glass transition on the mechanical and optical properties of freeze-dried apple and banana slices.* Journal of Food Engineering, 2011. **106**(3): p. 212-219.

37. Oikonomopoulou, V.P., M.K. Krokida, and V.T. Karathanos, *Structural properties of freeze-dried rice.* Journal of Food Engineering, 2011. **107**(3-4): p. 326-333.

38. Sablani, S.S. and M.S. Rahman, *Pore formation in selected foods as a function of shelf temperature during freeze-drying.* Drying Technology, 2002. **20**(7): p. 1379-1391.

39. Chang, C.-H., et al., *Comparisons on the antioxidant properties of fresh, freeze-dried and hot-air-dried tomatoes.* Journal of Food Engineering, 2006. **77**(3): p. 478-485.

40. Duan, X., et al., *Trends in microwave-assisted freeze-drying of foods.* Drying Technology, 2010. **28**(4): p. 444-453.

41. Shishehgarha, F., J. Makhlouf, and C. Ratti, *Freeze-drying characteristics of strawberries.* Drying technology, 2002. **20**(1): p. 131-145.

42. Krokida, M., V. Karathanos, and Z. Maroulis, *Effect of freeze-drying conditions on shrinkage and porosity of dehydrated agricultural products.* Journal of Food Engineering, 1998. **35**(4): p. 369-380.

43. Awan, J.A., *Elements of Food and Nutrition.* Vol. 1. 2018, Faisalabad: Unitech Communications. 7-10.

44. Syed, M.I.u.R.a.Q.A., *Post-Harvest Technology* in *Handbook of Food Science and Technology*, I.A.K.M. Farooq, Editor. 2017, Univerity of Agriculture Faisalabad: Univerity of Agriculture Faisalabad. p. 263-279.

45. Dainty, R., *Chemical/biochemical detection of spoilage.* International journal of food microbiology, 1996. **33**(1): p. 19-33.

46. Awan, J.A., *Food Science and Technology.* Vol. 1. 2018, Faisalabad: Unitech Communications. 79-82.

47. Kumar, C., M. Karim, and M.U. Joardder, *Intermittent drying of food products: A critical review.* Journal of Food Engineering, 2014. **121**: p. 48-57.

48. Shofian, N.M., et al., *Effect of freeze-drying on the antioxidant compounds and antioxidant activity of selected tropical fruits.* International Journal of molecular sciences, 2011. **12**(7): p. 4678-4692.

49. MacKenzie, A., *The Physico-chemical basis for the freeze-drying process.* Developments in biological standardization, 1976. **36**: p. 51-67.

50. Tarafdar, A., et al., *Optimization of freeze-drying process parameters for qualitative evaluation of button mushroom (Agaricus bisporus) using response surface methodology.* Journal of Food Quality, 2017. **2017**.

51. Koc, B., I. Eren, and F.K. Ertekin, *Modelling bulk density, porosity and shrinkage of quince during drying: The effect of drying method.* Journal of Food Engineering, 2008. **85**(3): p. 340-349.

52. Earle, R.L., *Unit operations in food processing.* 2013: Elsevier.

53. Comeau, M.A., *New topics in food engineering.* 2011: Nova Science Publishers.

54. Banga, J.R. and R.P. Singh, *Optimization of air drying of foods.* Journal of Food Engineering, 1994. **23**(2): p. 189-211.

55. 周明辉 and 荚荣, *Plackett-Burman Design 与均匀设计法优化玫瑰色微球菌固定化脱氮的性能.* 微生物学通报, 2015. **42**(9): p. 1671-1678.

56. Ahuja, S., G. Ferreira, and A. Moreira, *Application of Plackett-Burman design and response surface methodology to achieve exponential growth for aggregated shipworm bacterium.* Biotechnology and bioengineering, 2004. **85**(6): p. 666-675.

57. Plackett, R.L. and J.P. Burman, *The design of optimum multifactorial experiments.* Biometrika, 1946: p. 305-325.

58. Vanaja, K. and R. Shobha Rani, *Design of experiments: concept and applications of Plackett Burman*

*design.* Clinical research and regulatory affairs, 2007. **24**(1): p. 1-23.

59. Fahloul, D., et al., *Effect of osmotic dehydration on the freeze-drying kinetics of apricots.* Journal of Food, Agriculture & Environment, 2009. **7**(2): p. 117-121.

60. Salazar, N.A., et al., *Mango (Mangifera indica L.) Lyophilization Using Vacuum-Induced Freezing.* World Academy of Science, Engineering and Technology, 2015. **9**(10): p. 566-570.

61. Pikal, M.J. and S. Shah, *The collapse temperature in freeze-drying: Dependence on measurement methodology and rate of water removal from the glassy phase.* International Journal of Pharmaceutics, 1990. **62**(2-3): p. 165-186.

62. Patel, S.M., T. Doen, and M.J. Pikal, *Determination of endpoint of primary drying in freeze-drying process control.* Aaps Pharmscitech, 2010. **11**(1): p. 73-84.

63. Barbosa-Cánovas, G.V. and H. Vega-Mercado, *Dehydration of foods*. 1996: Springer Science & Business Media.

64. Mellor, J.D., *Fundamentals of freeze-drying*. 1978: Academic Press Inc.(London) Ltd.

65. Mujumdar, A.S., *Handbook of Industrial Drying, revised and expanded*. Vol. 1. 1995: CRC Press.

66. Liapis, A.I. and R. Bruttini, *11 Freeze Drying*. Handbook of Industrial Drying, 2014: p. 259.

67. Ellab, *The Freeze Drying Theory and Process Things to Consider*. 2018, Ellab A/S: Denmark. p. 5-11.

68. Kaewdam, S., et al., *Mathematical model of freeze-drying on mango*. Journal of Agricultural Research and Extension, 2013. **30**(3, Suppl.): p. 56-67.

69. Ahmed, J. and S. Rahman, *Handbook of food process design*. 2012: John Wiley & Sons.

70. Babić, J., M.J. Cantalejo, and C. Arroqui, *The effects of freeze-drying process parameters on Broiler chicken breast*

*meat.* LWT-Food Science and Technology, 2009. **42**(8): p. 1325-1334.

71. Mayeresse, Y., et al., *Freeze-drying process monitoring using a cold plasma ionization device.* PDA journal of pharmaceutical science and technology, 2007. **61**(3): p. 160.

72. Hammami, C. and F. René, *Determination of freeze-drying process variables for strawberries.* Journal of Food Engineering, 1997. **32**(2): p. 133-154.

73. Khuri, A.I. and S. Mukhopadhyay, *Response surface methodology.* Wiley Interdisciplinary Reviews: Computational Statistics, 2010. **2**(2): p. 128-149.

74. Levi, G. and M. Karel, *Volumetric shrinkage (collapse) in freeze-dried carbohydrates above their glass transition temperature.* Food Research International, 1995. **28**(2): p. 145-151.

75. Gordon, M. and J. Taylor, *Journal of Applied.* Chemistry, 1952. **2**(9): p. 493.

76. Barbosa-Canovas, G.V. and A. Ibarz, *Unit operations in food engineering*. 2002: CRC Press.

77. Wang, N. and J. Brennan, *Thermal conductivity of potato as a function of moisture content*. Journal of food engineering, 1992. **17**(2): p. 153-160.

78. Progelhof, R., J. Throne, and R. Ruetsch, *Methods for predicting the thermal conductivity of composite systems: a review*. Polymer Engineering & Science, 1976. **16**(9): p. 615-625.

79. Cheng, S.C. and R. Vachon, *The prediction of the thermal conductivity of two and three-phase solid heterogeneous mixtures*. International Journal of Heat and Mass Transfer, 1969. **12**(3): p. 249-264.

80. Singh, R.P. and A. Sarkar, *Thermal properties of frozen foods*, in *Engineering Properties of Foods, Third Edition*. 2014, CRC Press. p. 197-230.

81. Choi, Y. and M.R. Okos, *The thermal properties of tomato juice concentrates*. Transactions of the ASAE, 1983. **26**(1): p. 305-0311.

82. Sahin, S. and S.G. Sumnu, *Physical properties of foods.* 2006: Springer Science & Business Media.

83. Siebel, J., *Specific heat of various products.* Ice & Refrig., 1892. **2**: p. 256.

84. Tavman, S., I. Tavman, and S. Evcin, *Measurement of thermal diffusivity of granular food materials.* International communications in heat and mass transfer, 1997. **24**(7): p. 945-953.

85. Renaud, T., et al., *Thermal properties of model foods in the frozen state.* Journal of Food Engineering, 1992.

86. Gordon, C. and S. Thorne, *Determination of the thermal diffusivity of foods from temperature measurements during cooling.* Journal of food engineering, 1990. **11**(2): p. 133-145.

87. Choi, Y. and M. Okos, *Physical and chemical properties of food.* American Society of Agricultural Engineers, St. Joseph, MI, 1986.

88. Jiang, H., M. Zhang, and A.S. Mujumdar, *Physico-chemical changes during different stages of MFD/FD*

*banana chips.* Journal of Food Engineering, 2010. **101**(2): p. 140-145.

89. Witkiewicz, K. and J. Nastaj, *Simulation strategies in mathematical modelling of microwave heating in the freeze-drying process.* Drying Technology, 2010. **28**(8): p. 1001-1012.

90. Fan, K., M. Zhang, and A.S. Mujumdar, *Recent developments in high efficient freeze-drying of fruits and vegetables assisted by microwave: A review.* Critical reviews in food science and nutrition, 2018: p. 1-10.

91. Duan, X., et al., *Browning behaviour of button mushrooms during microwave freeze-drying.* Drying technology, 2016. **34**(11): p. 1373-1379.

92. Huang, J. and M. Zhang, *Effect of three drying methods on the drying characteristics and quality of okra.* Drying technology, 2016. **34**(8): p. 900-911.

93. George, J.P. and A. Datta, *Development and validation of heat and mass transfer models for freeze-drying of*

*vegetable slices.* Journal of food engineering, 2002. **52**(1): p. 89-93.

94. Claussen, I.C., et al., *Atmospheric freeze-drying—Modeling and simulation of a tunnel dryer.* Drying Technology, 2007. **25**(12): p. 1959-1965.

95. Heldman, D. and G. Hohner, *An analysis of atmospheric freeze-drying.* Journal of Food Science, 1974. **39**(1): p. 147-155.

96. Wolff, E. and H. Gibert, *Atmospheric freeze-drying part 1: Design, experimental investigation and energy-saving advantages.* Drying Technology, 1990. **8**(2): p. 385-404.

97. Kuu, W.-Y., J. McShane, and J. Wong, *Determination of mass transfer coefficients during freeze-drying using modelling and parameter estimation techniques.* International journal of pharmaceutics, 1995. **124**(2): p. 241-252.

98. Wang, W. and G. Chen, *Heat and mass transfer model of dielectric-material-assisted microwave freeze-drying of*

*skim milk with hygroscopic effect.* Chemical Engineering Science, 2005. **60**(23): p. 6542-6550.

99. Luikov, A.V., *Systems of differential equations of heat and mass transfer in capillary-porous bodies.* International Journal of Heat and mass transfer, 1975. **18**(1): p. 1-14.

100. Ryan, D., R. Carbonell, and S. Whitaker. *A theory of diffusion and reaction in porous media.* in *AIChE Symp. Ser.* 1981.

101. Whitaker, S. and W.T. Chou, *Drying granular porous media-theory and experiment.* Drying technology, 1983. **1**(1): p. 3-33.

102. Crapiste, G., S. Whitaker, and E. Rotstein, *Drying of cellular material—I. A mass transfer theory.* Chemical Engineering Science, 1988. **43**(11): p. 2919-2928.

103. Nastaj, J. and K. Witkiewicz, *Mathematical modelling of the primary and secondary vacuum freeze-drying of random solids at microwave heating.* International

Journal of Heat and Mass Transfer, 2009. **52**(21-22): p. 4796-4806.

104. Erdoğdu, F., *A review on simultaneous determination of thermal diffusivity and heat transfer coefficient.* Journal of Food Engineering, 2008. **86**(3): p. 453-459.

105. Arévalo-Pinedo, A. and F.E. Murr, *Kinetics of vacuum drying of pumpkin (Cucurbita maxima): modelling with shrinkage.* Journal of Food Engineering, 2006. **76**(4): p. 562-567.

106. Delgado, A. and D.-W. Sun, *Heat and mass transfer models for predicting freezing processes—a review.* Journal of Food Engineering, 2001. **47**(3): p. 157-174.

107. Brülls, M. and A. Rasmuson, *Heat transfer in vial lyophilization.* International Journal of Pharmaceutics, 2002. **246**(1-2): p. 1-16.

108. Vega-Gálvez, A., et al., *Mathematical modelling of mass transfer during the rehydration process of Aloe vera (Aloe barbadensis Miller).* Food and Bioproducts Processing, 2009. **87**(4): p. 254-260.

109. Santacatalina, J., et al., *Model-based investigation into atmospheric freeze-drying assisted by power ultrasound.* Journal of Food Engineering, 2015. **151**: p. 7-15.

## ABOUT THE AUTHOR

Muhammad Waseem Akbar is a food engineer by education. He is expert in nutritional, health and processing courses. Drying is a preservation technique. He has a special interest in the freeze-drying technique. This is one of the most emerging drying technique and has the potential to produce quality products with an extended shelf life for years. The author has written multiple books in the food science and engineering domain

www.ingramcontent.com/pod-product-compliance
Lightning Source LLC
Chambersburg PA
CBHW070446220526
45466CB00004B/1774